May the Lord Have "MRSA" On me!

This is a short, amusing, yet very informative story of my unbelievable and extremely painful experience with the super-bug/bacteria known as; **"MRSA".** This is a flesh eating bacterium that has changed my entire view of the bacteria which surround me today. And now I would like to share with you, my educational narrative of how I was able to combat this vicious, man-eating monster, which devours human flesh; hunt it down and kill it! And, with the help of the professional staff, at "**St. Anthony hospital",** in Lakewood, Colorado, I was able to conquer this evil adversary with total and complete victory. I am now known as; **"The-MRSA-Nator"**

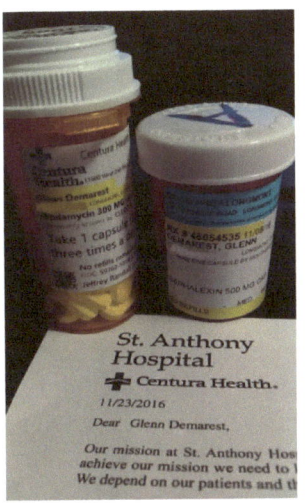

Glenn Michael Demarest
2016

It was a Sunday morning, November 6th, in the year of 2016. And it was the warmest and sunniest day in November that I think I have ever seen on any given day of that month because normally our temperatures here in Colorado this time of the year, are far much cooler by November. Anyway, I awoke that Sunday morning with the symptoms of what I thought to be the beginnings of an average head cold. And by the end of that very memorable Sunday it became even more severe than I would have ever anticipated. This thing really took me by complete surprise and I had absolutely no idea that it would literally kick the living crap right out of me.

   I mean it honestly made me feel like I had been in a boxing match and not only lost the match, but was totally slaughtered in that match while I strategically blocked every single incoming punch with my face; and especially with my poor unsuspecting nose.

☹ Monday morning came around and I got out of the bed and went straight to the shower and took a nice, long, and hot one. While letting the water just beat down upon my nose, my face and my forehead, I noticed the top of my skull was feeling extremely sore as if someone hit me right on the top of the head with a wooden spoon.

   And the only reason I describe it to feel like that is because "believe it or not" I remember being hit on the top of my head with a wooden spoon by my mother when I was a small boy. It was her little unique and old-school way of saying "behave"! And if you needed any reminder after that you were taken to the next level which would be the "ping-pong-paddle" from the table we kept in our basement next to the laundry room; and she would whack me right on the bare bottom. Yes, things in life were much different back then.

   Anyway, after my shower I got dressed and drove to the nearest pharmacy to purchase the strongest head cold medicine that they had, and when I returned to my home I took that behind the counter medicine immediately and then changed into my sweats, turned on the flat screen and mounted the couch with my old and trusted friend "blanky" as I waited for my soup and tea which I had just previously prepared in the microwave oven.

I really didn't think anything about my head cold and I didn't have to go to work the following Monday morning either, so to me this was simply going to be a couple of days off in front of the flat screen with the entire house to myself, Yay! Well, that pleasant thought of a couple of days off was to be very short lived and I will have soon learned yet another one of my life's most painfully unforgettable lessons. Armed with my "Sudafed, my tea with honey, and my bottle of Nyquil", all of which lay upon the table next to me at my bedside.

A bottle of water in one hand and the remote control in the other hand, I went to bed that night thinking what everyone else in America would think. Okay; no big deal. In a couple of days I'll be fine. I turned the lights out, and then that was that and I simply fell asleep. Monday morning I had awoke at four-thirty in the morning with an extremely strong urge to urinate very, very, badly! It felt to me like I had been holding it for several hours or more and that I was about go pee right at that very moment, and then I realized that I had a horrible case of the old "dry as a bone" cotton mouth as well.

I sat up in the bed and immediately became dizzy, then with what felt like a hangover I staggered into the master bathroom, turning on the light switch as I stepped across the threshold that separated the bedroom carpet from the Pacific pattern floor tile which I had personally installed the year before. I saw my reflection within the bathroom mirror that faced me and saw what actually appeared to me at that time to be what I can only describe as some kind of a "human cartoon". I looked totally hideous!

☹ In fact my first reaction was simply, "wtf"? And then it was more like a **"WTF"?** And then after that, of course, it was more like **HOLY SHIT! OMG! WTF!** I realized at that very moment I was suffering from more than a head cold and quite possibly a serious sinus infection. ☹ Okay, relax, don't freak out I told myself as I already have an appointment with an "ENT" Doctor (ear, nose, and throat) on Tuesday morning, so to me there was no reason to call him or any other Doctor at that time; so I didn't. And so by bedtime that night I was hurting pretty bad and I then took a sleeping pill and an extra pull or two from the Nyquil bottle, crossed my fingers, closed my eyes and wished to myself that morning might actually bring some relief; I was wrong.

There is absolutely no relief with "MRSA", and I can absolutely assure you that there is absolutely no mercy with **"MRSA"** either. So, there I was comfortably couched and magnificently mounted to my sofa, laying there watching the History Channel and waiting for yet one more day to pass so I can finally see the Doctor and go home with some kind of prescribed comfort in a capsule which will hopefully make my life a happy life again. That day was an extremely long day for some reason, and nothing makes a clock move slower than starring at one; trust me.

But somehow I managed to get through that day as well as that night and perhaps it was the extra shooters of Nyquil, I don't know but I made it to the next morning. But when I woke up again at five-thirty in the morning, I was horrified to see what I looked like when I saw my ugly reflection in the mirror. It was like looking into the fun-house mirror at the carnival or county fair. Not only was the pain in my face a fierce and burning pain, but my whole face was swollen and wrinkled like an oversized Mr. Potato-head turned pumpkin with that wrinkled prune thing going on.

    I looked like something that was right out of a horror movie, and when my sixteen year old step-daughter looked at me and so kindly expressed that I looked like a ninety-nine year old man, it made me cry. ☹ Then I suddenly realized that I was truly in trouble and that this was something which I have never seen before. So, with great effort, I got into the hot shower once again and began my long and grueling day which would soon be followed by the hour long drive to the "ENT" Doctor in Denver.

    Even after a long hot shower I still looked like shit on the pavement after rush hour had passed over it ten-thousand times and physically I felt even worse. And I also knew at that point in time that my so called "couple of days off" was now going to be at least a week of pain and suffering. Now I can afford a couple of days off from work, but a whole week off? Now that's a different story altogether and I just can't afford that.

    After all, I have an ex-wife, a step-daughter, who I refer to as "my daughter" and rent and bills to pay. So like I really need to go to work dude, or else I will simply go insane! Ya know what I'm saying? Being with my daughter is my number one priority. Being at work is my number two, and after that I am usually not very happy due to so many health problems that I have accumulated these last two years.

    Well, before I knew it, the time had passed and it was now time to go to my appointment.

    I got into the car and began my one hour drive to Denver to see this "ENT" who I was actually scheduled to see for a tinnitus exam as I have also been having trouble with my hearing lately, so he has no idea that I am coming down there to share with him my big, red, wrinkled pumpkin-head and swollen strawberry nose.

And I hope to dear "God" that he can fix this embarrassing Halloween pumpkin face that I am now sporting. My head literally looks like an old, swollen, wrinkled pumpkin with a sporty, swollen nose that looks remarkably like some kind of a large and giant, angry strawberry gone rouge and on the rampage while high on "crack" or something. And I will tell you right now my friend; I got some of the most paralyzing looks that I have ever seen from people as they peered through their automobile windows at me while I was driving.

I thought to myself. Thank god I'm not a disfigured leper with "Elephant Titus" or something. I can only imagine how they would look upon me then. It made me realize right there how cruel people can actually be, especially adults; wow... As if I didn't already feel bad enough, I wish I had a ski mask. Or better yet, the masks from the movie scream that would have attracted far less attention. ☺

I got more stupid looks from the adults then I did from the children who saw me, and to be honest with you I would have expected to see just the opposite, funny how that works. Grown adults running around and acting like children, no way; really? Well, I had finally made it into the parking area of the Doctors office and found a convenient parking space to park my little economy car into.

I waited a minute for some of the other people I saw entering the building to disappear as so I can enter un-noticed as I had grown tired of stupid people looking at me all-stupid like, as if I was some kind of "circus freak" like you see in the old black and white movies on **"Sven-Ghoulie Night"** every Saturday evening on "ME-TV". Anyway, I got into the Doctors office and finally two people who were sitting behind the reception counter were looking at me, and not my big-red-swollen-face!

And then I began filling out the paperwork. A door to my left soon opened and a voice called and asked me to step inside. After the assistant performed the exam, the Doctor entered the room to look at my wrinkled, swollen, pumpkin-face and the world largest upside-down strawberry which was now attached to it; and he got to see all of this with no cover charge!

He began telling me how handsome, in shape and muscular I was looking and that I should truly consider dating his sexy, beautiful, swimsuit-model daughter who was only twenty-four years old.

And then I woke up. Lol… Feeling dizzy, light-headed, and stupid. Because I let myself slip into some weird and crazy dream. This illness has definitely affected my mind. He began explaining to me that he was prescribing me "Cephalexin 500 mg", and that I should take one capsule every six hours for the next ten days.

Then, as he was walking out of the door he suddenly stopped and turned around and said to me, if your condition does not improve within the next three to four days, I think that you should get yourself to the nearest hospital emergency room.

Okay Doc, Thank you! I said to him, and I left the building and drove myself back to my home in Longmont, or "Wrong Mont" as I like to call it.
"Attention ladies and gentlemen; Pumpkin-Head has left the building"! The drive home was a little spooky to say the least as I was feeling very light-headed if not downright dizzy, and I could not breathe at all through my nose as it was literally swollen shut, and that alone was a total distraction.

My god; I hate being a mouth-breather. All in all I was more than likely in no shape what so ever to drive. But, since I had no choice and no one else to drive me, I drove myself, what can I say; I'm old school, and hard-core. When I arrived at my home, I realized that it was time for me to just climb back into my bed and rest, something that I don't get to do very often because I work almost seven days a week and almost every week. So I never seem to get enough sleep, because I'm either working too much or I am physically hurting so badly that I can't fall asleep.

After taking my prescribed medication along with all of the other medicines that my regular "primary care physician" has me on for all of the other health problems I have been experiencing these last two years, I then took a sleeping pill in the hopes of just sleeping for the next two or three days.

After all, I have been told a hundred times by many medical professionals that the human body heals best when one is sleeping. So that was my game plan, sleep, take pills, smoke some weed and then sleep some more. ☺ I was hoping that after two to three days I would wake up and notice a significant change in my current health condition and I can then get back to the daily grindstone of reality, simply meaning get back to my work. One thing I learned a long time ago, being at home too much can be dangerous for a guy like me.

Because working is very therapeutic for me, and sitting around with nothing to do but watch daytime TV makes me realize how stupid the majority of the American public has truly become. And that tends to irritate me and make me somewhat short tempered with all of the idiots when I am out in public. The next thing I knew it was now morning and it was time to get up and take my "meds".

I was also feeling hungry which I thought was a good sign. However, my face hurt so bad that I just could not chew any food without experiencing severe pain as I did so. So I made some cream of chicken soup that had nothing in it that needed chewing, and that made it easy to swallow. I dunked some bread into the soup to add as filler and fiber and then drank a glass of milk with it, and having something in my belly brought a feeling of improvement within minutes.

Then I made myself a cup of green tea which I drank and choked on as I watched the morning news. After a couple of hours I took another sleeping pill and I fell fast asleep hoping that I could sleep away yet another day of facial misery.

This was my daily routine for the next three days, and before I knew it, it was now Friday night and I had noticed that my condition had only worsened. That night I came to the conclusion that I was definitely in serious trouble and that it was time for me to get my sickly butt to the "Longmont United Hospital" emergency room to get looked at.

I decided that it would be best for me to rough it out for one more night and go to the hospital first thing in the morning. That would be the best plan for me in my opinion; because my daughter needs me to drive to her work at ten-o'clock in the morning as well. So, I will drop her off at work and then drive myself to the emergency room and who knows, maybe when I wake up in the morning I might actually feel a little better.

That was my second mistake as I was stupidly hoping that the antibiotics would show some sign of improvement and old wrinkled pumpkin face with the giant "strawberry nose" would not only feel a little better but look a little better as well. Unfortunately for me, and as much as I hate to admit to it, I was once again wrong.

Seems like I have been saying that a lot these past two years, and I'm getting really tired of it. That Saturday morning I had awaken to some of the most horrible facial sinus pain and pressure that I have ever experienced in my fifty-three years of life on this crazy planet, and then I noticed that my pillow was soaking wet with tears and blood.

And for the life of me I could not get the tears to stop running like the faucet of a sink that was left slightly on. My big fat, ugly swollen, wrinkled pumpkin face was so swelled up that it was literally pushing the tears right out of my tear ducts, and it just wouldn't stop!

And now I felt almost helpless, like a little child; it was nothing short of just plain freaking horrible!
At nine-forty in the morning, my daughter and I walked out of the house and got into my little ford, two-door car. We then pulled away from the curb and down the street we drove.

On the way to her job we both sat in total silence. She could tell that I was in serious pain and I know that she could see that I was crying on top of that, and when she ask if I was alright, I turned my head towards her and softly said no honey-bunny; I am not.

☹ Suddenly I realized that the turn into the parking lot of the store that she works at lay before me just ahead and to the right. I pulled in and stopped the car at the front entrance of the store, touched her hand with mine, and said to her, "I love you" honey-bunny", which is the affectionate nick name I gave to her a long time ago.

Because at that moment, my gut-feeling inside, told me that whatever was ailing me, was going to kill me and take my life at any moment. That's how bad the pain in my face was. I have pretty much seen it all. You name it and I have broken it, fractured it, shattered it, or been infected by it.

But whatever this was; it was something I have never seen before. And I don't want to leave this world without saying good-bye and I love you to my sweet-little daughter; she is the only thing I've ever done right; and the only thing I have left in this world to prove it.

My beautiful daughter. "Suhaira".

  I had that happen to me once before when I lost one of my older brothers to liver cancer and I never got to say good-bye to him and tell him that I loved him because he died so quickly. And I swore from that moment on, that I would never let that happen to me again. I look back upon that moment now and realize that the reason she didn't answer me was probably due to the fact that she was more scared and freaked out than I was.

  After all, I really was hideous and monstrous looking; hell it even made me cry when I looked into the mirror. As I drove away and merged onto the highway towards the Hospital in Longmont It began to register how painful this thing in my nose had become, and that the tears were flowing non-stop and dribbling like crazy.

  Oh; my; gosh! I would have never guessed that we humans stored so many ounces of tears in us. ☹ I mean this stuff was freely flowing on tap today, and the tap was stuck open in the pour position. Did you know that in healthy "mammalian" eyes the cornea is continually kept wet and nourished by basil tears?

  They lubricate the eye and help to keep it clear of dust. Tear fluid contains water, mucin, lipids, lysozyme, lactoferrin, lipocalin, lacritin, immunoglobulins, glucose, urea, sodium, and potassium? Well, don't feel stupid or anything because I didn't know any of that stuff either until I began doing some research to write this book. The film of your tears by the way is made up of three layers including oil, water and mucus. The water layer is the thickest layer and it contains nutrients that are crucial for your eyes.

The oil layer reduces evaporation to keep the eyes moist. Some researchers have also suggested that emotional tears contain stress hormones, which the body is able to physically push out through the process of crying. A study that I read says that the average person can shed up to ten ounces of tears per day, and up to thirty gallons a year. Wow! Who in the hell is doing all of that crying out there; for crying-out loud! Ha-ha, I'm so funny!

☺ I think that it is time to give that loser the boot and find someone who makes you laugh instead of making you cry! They say that crying is good for you, that maybe so but I know first-hand that laughter is better. Hence the old saying, laughter is good medicine; and it is trust me! I know this to be true. I have laughed my way through many injuries, especially once I realized how stupid it was, or should I say; how stupid I was to allow myself to become so stupidly injured. Okay, let's get back to my story now.

So, I finally got to the emergency room entrance in the hospital parking lot, and lucky me I found a parking spot that was less than five-hundred and fifty feet away. I grabbed only what I essentially needed and walked inside. Everybody looked at me with their mouths wide open as if my swollen face was about to make them vomit. Wow, thanks folks, as if I didn't feel bad enough as it was. Thank goodness the staff didn't look at me like that, and they treated me very kindly and quite respectfully by the way.

And they even called me by my name, instead of **"Pumpkin-Head"**, or **"Wrinkle-Face"**. I entered the "ER" immediately and they quickly put me into one of the rooms and began the usual stuff like blood pressure, temperature, and lots of questions like; do you have any allergies?

Just marriage, ex-wives, and stupid people who act like simple minded live-stock; I replied. Nobody laughed but me, and then I realized how much it hurts to laugh; and then I felt stupid again. But I also noticed that I was the only person who seemed to know anything about marriage or divorce; so maybe I just simply know something that they don't know yet.

I guess I just didn't realize the severity of my condition, as I still thought that I was just suffering from a sever sinus infection, or at least that's what I was hoping to hear. I would have never guessed in a hundred years that there was a flesh-eating monster inside of my nose just chowing down like this was some-kind of all you can eat buffet and open twenty-four hours a day.

And not only that the little bastard didn't even pay to get in, and absolutely no gratuity. Son of a bitch! I hate when that happens! ☹ Finally a Doctor came into my room and began asking me more questions while he poked around and peered into my giant, upside down, strawberry nosey-rosy.

Does it hurt when I do this? Uh, yeah, I replied. Did you think that it would feel good to me? On a scale of one to ten what would you rate your pain level at? Um, a twenty-five and some change, I said to him. They immediately inserted an IV into my arm and told me that this will help you with the pain, and that I should just relax while they take my blood.

And we're also going to administer some antibiotics and something for your pain into your arm as well, do you feel okay?

Well, I feel better just being here, but I still feel like shit on a hot-highway in July and I'm still pretty freaked out as to what it is that is doing this to my face.

They finished taking my blood and ran it to the lab. Then, they gave me a "C-T" scan and returned me to my room. Now I don't know how long I was lying there, but when the Doctor returned and began explaining to me that I had a **"MRSA"** abscess inside of my nose that was trying to work its way up to my brain, I was like; **"WTF"?**

The Doctor then began to inform me that I was going to be transferred to another hospital where I can be seen by an "ENT" specialist who is going to perform emergency surgery immediately. And that the hospital I was being taken to was **"St. Anthony Hospital"** in Lakewood just south west of downtown Denver.

Are you kidding me I asked? Don't you have one of those things here? An "ENT" is an ear, nose, and throat specialist, and it's not a thing, it's a Doctor. I don't care if it's a freakin monkey, don't you have one here? And if not, why not, and why don't you keep one here? We are unable to get our hands on one of those things today Glenn. Okay, I'm with you, I get it. And once again I felt kind of stupid, and I immediately texted my daughter as to where I would be going next.

I kind of figured that I had something really bad inside of me, but I guess I just didn't want to worry about it too much until I absolutely had too, ya know what I mean? Well, I had to lay there and wait for the ambulance that was coming up from Denver to transport me, and the staff was checking on me very often to make sure that I was comfortable, so I have to say that they did a very good job at comforting me and making sure that I wasn't suffering.

And when the ambulance arrived it was about seven-o'clock in the evening, Damned, I thought to myself; we are going to be right in the middle of evening traffic on a Saturday night. But that wasn't the worst of it. There was road construction everywhere, and on top of all that, all traffic had come to a complete stop because of a two car accident somewhere on I-25 by the Thornton Parkway. Wow... My luck just keeps getting better and better this week.

I'm on a roll, baby! When the two ambulance guys arrived we quickly made our get away without paying my tab. Ha-ha! One of the ambulance guys had an app on his phone that shows you the accident and all of the details with it, that's pretty cool I said to him. Those guys were great.

They made sure that I was comfortable, warm, and then they started jamming some old school **"Led-Zeppelin"** on the stereo which I thought was awesome by the way. And they truly did make the best of what was a really shitty situation for me. The drive to Lakewood seemed to take forever, and just as we arrived I was about to become tired of bouncing about in the back of an ambulance. The guys rolled me into the hospital and up to the sixth floor and then into room number #607.

We said our goodbyes and then parted ways as they probably had another fare to pick up someplace elsewhere. And the next thing I knew there was a whole bunch of other hospital staff members entering my new room with the big giant six foot flat screen "TV". Okay, okay, it wasn't that big but it sure looked like it was that big when I was all doped up on that happy stuff they put in my arm through the "IV".

I can't remember who I met first and it seemed like every one was all talking at the exact same time, and it also sounded like they were talking through a large drainage pipe.

And as a construction kind of guy, that is the best way that I can describe it to you. Every voice just seemed to have an odd "echo" to it. I don't really remember that first night much, I just remember what seemed like a whole bunch of people coming into my room to see me and fuss over me; and I liked it. I don't get treated like this at home, that's for sure.

Then, someone explained to me that I would be undergoing surgery at six o'clock in the morning and a bunch of other stuff that I should know but can't remember because I was jacked-up on shots of pain medicine in my IV drip line.

I was also really, really tired, and don't forget that this condition made it impossible for me to eat any solid foods for days, only liquids; so I was feeling very weak as well. And after that the next thing that I can remember was just falling asleep. It felt like only minutes had passed when I was awakened by what looked like two people wearing space suits abducting me and then they took me to a secret room deep inside of the "Mothership".

Wow, I said out-loud; what a trip dude? And all of a sudden I could hear a very old song that was popular in the year of 1966. It was a song by a then popular rock and roll band known as the **"Byrd's"** and the song was titled **"Mr. Spaceman"** and all of a sudden I felt very comfortable, and extremely mellow…. I also came to the realization that I was in the "O.R." and that these are the folks who are going to perform my surgery, and that they are not space-men.

Then it all came back to me and I remembered the Doctor, surgeon, "ENT" guy, and he seemed like a real cool dude so I once again felt very comfortable and at peace with the situation. I was no longer freaked out over this crazy, flesh eating monster that was hiding inside of my nose, and I suddenly realized that everything was going to get better now and that I was going to be ok. I can still see the C-T scan image in my mind which looked like some kind of cartoon monster in my nose.

And the next thing that I remember was the anesthesiologist person informing me that I was going to go to sleep now or something like that; and then I did. That was it, and that is really all that I can remember and then I awoke some hours later back in my hospital room, penthouse suite number #607. It had a big flat screen TV and the big-booming theater surround-sound system and the flowers, the balloons and the dancing girls who were all dressed up like nurses. We were having a big-ole party! Oh, and the nurses! They were all so beautiful! Even the hairy one was gorgeous!

Oh, whoops! I guess that nurse was a dude. And then once again I woke up only to realize that I had been dreaming all of this crazy stuff all of this time and that I was in the hospital and not in the most exciting, weirdest, and exotic dream that I have had in well over sixteen years or possibly more. ☹ After several minutes passed by I was able to come to my senses and I immediately noticed an enormous difference in the pain and the swelling and that I was actually able to slightly breathe through my poor nose once again.

And then I could feel that my nose was a good two to three sizes smaller than that triple-x, upside down, giant strawberry which more than likely could have passed for a hot-air balloon without the basket attached below. I found my cell-phone on the tray-table next to my bed and took a selfie so I could see what I looked like. Wow! I'm human again! It was such a relief to see myself looking healthier than I had been looking this last week, and I immediately felt ten times better as well.

Not to mention the incredible and comfortable body-numbing-buzz that I had going on. To this day I still have no idea what they were pumping into my arm through that "I.V." drip-line but it sure was some good stuff. Suddenly, the door to my new "Penthouse suite" burst open and all kinds of wonderful people came flooding into my room one by one and one after another; they just kept coming! It was like we were having a big party and everyone was so glad to see me; it was awesome!

I felt like I was some kind of "Hollywood celebrity" and then I met a really nice guy who said that he was my doctor, and then I met this beautiful nurse who said that she was my wife, (just kidding honey) my wife thinks I'm crazy; so does my daughter. Oh yeah, then I met that cool dude from before in the "ER", and he told me that he was my surgeon and that everything went well. And then there was that nice lady who said the she would be my personal "on-call" twenty-four hour a day "masseuse", and I'm not sure but I could have sworn that I saw a limo driver with a gloss-black stretched limo parked back by the rest room; man I was having the time of my life!

They even had a lady wearing a bikini cleaning my swimming pool. It was a giant crowd which now surrounded me; it was a magnificent moment that brought a tear to each of my eyes. Then there seemed to be a whole bunch of other people there too, I couldn't see them because they were toward the end of the line which went out of the doorway and out into the hallway.

I think they even had a couple of big, stocky, muscle guys working the door, taking a cover-charge and checking "ID's". It was truly amazing and un-freak in-believable! And then like (POW)! Something sucker-punched me right in the back of my head forcing me to look downward, and as I lifted my head back upward everyone was gone, even the pool-lady. And standing before me was my surgeon, my med-doctor and my nurse for that shift.

And that is when I realized that once again I had been dreaming and then I felt kinda dizzy and stupid again. Now the entire room had become deafly quiet, and I could hear the "ENT" surgeon speaking in a low tone and in really, really, slow motion.

Then, I heard him say; and I quote. "The "MRSA" bacteria has completely eaten away all of the cartilage in your nose" and if you ever have another abscess like this in your nose again? Or if you ever get punched in the nose; your nose will more than likely collapse. And then you will seriously need to have reconstructive surgery because your nose is going to look like a "platypus". That right there was all he needed to say to me because my mind instantly went into "OMG" mode and initiated my automatic "shut everyone out mode" because I am now deep in thought and nothing is funny right now.

And after that, everybody just sounded like a faint echo on a windy day in the far distance as the severity and the reality of what has happened to me just hit me in the face, stomach, and in the ball-sack at the exact same time. And it made me feel sad as it took my breath completely away for a moment, either that or I just stopped breathing for a minute or so; I catch myself doing that sometimes and I don't know why.

And it was like a drop to my knees kind of sad because I for the life of me could not figure out what, when, or where I have been lately for this horrible, flesh eating monster to get up inside of my nose. And why did it pick my nose and not someone else's. "No pun intended". Why couldn't this have happened to someone who I don't like or something? ☹ Actually, I would not wish this horrendous illness upon my worst enemy; it really is that bad.

The next thing I remember after that was a beautiful nurse who was administering a syringe into my "IV" drip line and she said to me with a most pleasant smile, "wha-wha-wha? Wah-wah-wah-wah"? And then I said, Okay… But I have no recollection of what she actually said to me. And once again it was "lights-out" and I was fast asleep like an ittle-yittle, baby-boy. As I lay there in my warm bed and enjoying my much needed slumber, I began to dream again, and everyone around me kept asking me all kinds of weird and crazy questions.

Have you ever had "MRSA" before? Do you pick your nose, and if so how often? Are you practicing good hygiene and washing your hands frequently and scrubbing under your finger nails with a scrub brush? Are you or have you ever snorted any drugs up your nose? Do you change your underwear daily? Do you wear boxers or briefs? I just looked at them, shrugged my shoulders and replied; depends! Have you ever cheated on your taxes? "WTF" kinds of questions are these which you ask of me?

And of course I wash my hands frequently, every time I use the restroom, before every meal, regularly at work because my work is usually a dirty environment, and I shower every day. Furthermore, in response to your question about snorting anything up my nose; I haven't done anything like that in over twenty-five years or more.

But, if I am going to suffer like this; then I wish I had. At least then I would have some kind of reasoning behind all of this pain and suffering. And why did this shit decided to go for a hike and climb up into my nose and set up camp! Then, they began asking me even more questions; but the only one that I remember was the one about my pain level. Now, having my sinuses, my nose and my face so severely swollen like that seemed to affect my hearing quite a bit, and possibly my mind.

Because I kept thinking that they were talking about some type of construction tool or something. Hey, hand me that pain-level with that cordless catheter and the number two Philips-head "IV" pouch. It was so weird and crazy, like some kind of **"Rocky Horror"** bad dream picture show, and being a guy who has spent all of his life performing some type of construction; I guess my mind was relating to everything as if it were some type of tool on some type of a job site somewhere.

Suddenly everything in my mind came to a complete halt when two doctors entered my room wearing some type of protective plastic "cover-all" space-man looking outfit.

They immediately told me that I was now in isolation which is why they were wearing the "zip-lock" looking "hefty-bag costumes. One was an older and distinguished looking gentleman and he introduced himself as "Dr. So and so". And even if I could remember his name, I didn't get his permission to mention it in this book. So that right there is as good as that's going to get.

The other one was a tall and lovely looking blond woman with big beautiful blue eyes, and I'm not one-hundred percent sure "because I was so goofy on all of that pain medicine" but I could swear that she said her name was **"Dr. Hottie"**... And she was so nice and so sweet, and such a gorgeous looking "hottie" that I could not hear a single word that she was saying to me. All I could hear her say was; I love you Glenn!

And I want to make love to you all day and all night! And someday I will have your baby! Instantly I shook my head violently and came to my senses just in time to hear her ask me if I were in pain; and on a scale of one to ten how would I rate your pain level. I told her that she was a ten; I mean that the pain level was a ten. The other doctor began to speak again, but I just ignored him as he was not at all my type.

Besides, I was in an overwhelmingly enchanted trance by my new girlfriend **"Dr. Hottie"** and I just couldn't stop gazing at her big blue eyes. Even in that zip-lock baggie she looked hot! Suddenly and unexpectedly, I found myself lying on a white sand beach with my new found love **"Dr. Hottie"**. We were rubbing sun-tan lotion upon each other's golden, toned bodies, and she repeatedly told me how much she admired my six-pack abs and my muscular, muscleman build. I told her that we would be together forever and that I will never leave her side.

Then I told her how I would protect her from the "evil doers" and bring worldwide peace to this troubled planet. I shook my head again and slapped my face, really, really hard. But not in the nose! "Stop it and concentrate, damn it! I yelled to myself! It was so loud that for a moment there I thought that everyone else in the room could also hear what was going on inside of my big, swollen, pumpkin-head.

My god, I am such a dork, sometimes I even embarrass myself; really I do. I then suddenly felt mentally troubled. That's when I heard on the intercom out in the hallway. Paging **"Dr. Hottie"**; paging **"Dr. Hottie"**; Code blue on the balls please in room #607 **"STAT"**! My god Glenn; **"STOP-IT"**! I am so sorry! I am usually not this bad, okay, okay, maybe I am; but I'm trying really, really hard not to be.

Anyway, That's when I peeked under the covers and I could see that I was still wearing a hospital gown and below that I could see my skinny, pasty-white, anemic looking chicken legs as I was no longer lying on the beach and I then felt a tear swelling up in my eye as I lay there pouting.

☹ She then began explaining to me numerous things that I could do to ensure that **"MRSA"** does not sneak-up from behind and attack me like this again, she told me that I should wash my hands constantly, especially doing the kind of work that I do and how that is the first line of defense against spreading **"MRSA"** as its favorite place to hide is under ones finger nails. Scrub your hands with a scrub brush when possible for a minimum of 20 seconds before drying them with a clean towel, and use another towel to turn of the water faucet.

It's also recommended to carry hand sanitizer that contains sixty percent alcohol; this can keep your hands clean when you don't have access to antibacterial soap and water. If you have any wounds like cuts, scratches or rashes, keep them covered at all times. Covering wounds can prevent pus, blood or other fluids containing staph bacteria from contaminating surfaces that other people may touch.

Don't share any personal items; this includes towels, sheets, razors, trimmers and athletic equipment. People with **"MRSA"** are typically placed in isolation temporarily until there infection improves. Isolation prevents the spread of this type of **"MRSA"** infection. Hospital personnel caring for people with **"MRSA"** have to follow strict hand-washing procedures to further reduce their risk for **"MRSA"** and hospital staff and visitors are required to wear protective garments and gloves to prevent contact with any contaminated surfaces.

Linens, clothing and any contaminated surfaces should always be properly disinfected. **"MRSA"** can spread from a small contained infection to a larger one that involves your internal organs and body systems. It has been linked to pneumonia and blood stream infections like sepsis. Recent "CDC" reports found over 72,000 severe **"MRSA"** infections and over 9,000 deaths per year right here in the United States alone!

Most **"MRSA"** is associated with infections that are transmitted through close personal contact with an infected person or through direct contact with an infected wound. Many types of **"MRSA"** infections may also develop as a result of poor hygiene such as infrequent hand-washing. **"MRSA"** symptoms can vary depending on the type of infection, and any areas that have been cut, scratched, or excessively rubbed or any rashes are also vulnerable to infection because your biggest barrier to germs, your skin, has been damaged.

The infection usually causes a swollen, painful bump to form upon the skin. The bump may resemble a spider bite or a pimple. It often has a yellow or white colored center and a central head. This may often be surrounded by an area of redness and warmth, known as cellulitis.

Pus and other fluids may drain from the affected area and some people may also experience a fever. **What exactly is "MRSA"?** Well, the abbreviation of **"MRSA"**. Simply stands for (Methicillin Resistant Staphylococcus Aureus). And, it is one, "bad-ass", mother of all bacterium that does not play well with others in the sand-box. Nor does it respond well to most antibiotics. See the microscopic pictures below.

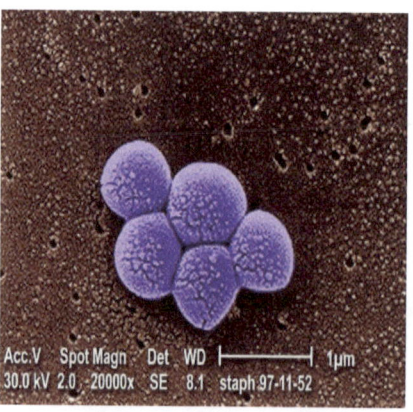

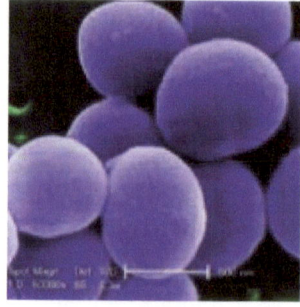

Though most infections are minor, some infections can be extremely life threatening. The **"MRSA"** infection, pronounced **(Mur-Sa)** by the way is an infection which was supposedly discovered in 1961 somewhere in the "UK" and supposedly appeared twenty-years later here in the United States in the year of 1981.

While I was doing some research on this barbaric "bad-boy" bacterium, I stumbled upon a conspiracy theory from a writer who believes that the **"MRSA"** bacteria was specifically designed by man to be implemented in the case scenario of "Biological warfare" as opposed to nuclear warfare to kill and incapacitate the enemy during the "cold war". Now that right there is some crazy-ass information that I wouldn't doubt for one moment.

I know, it's just a theory but when I was a young boy during that so called "cold-war" I remember some of the crazy and insane things that the United States and the Soviet Union were doing back then, and how they each were looking for new ways to kill each-other without engaging in nuclear or conventional warfare.

But we won't get into that right now, as I will save that for a different book and another time. **What are the different types of MRSA?** "MRSA" infections are classified as either hospital-acquired **(HA-MRSA)** or community-acquired **(CA-MRSA).** **"HA-MRSA"** is associated with infections that are contracted in medical facilities such as hospitals or nursing homes. You can get this type of **"MRSA"** infection through direct contact with an infected wound or contaminated hands and surfaces.

"CA-MRSA" is associated with infections that are transmitted through close personal contact with an infected person or through direct contact with an infected wound. This particular type of **"CA-MRSA"** infection may also develop as a result of poor hygiene such as infrequent and or improper hand-washing and it can be very contagious and can easily be spread through direct contact.

And though a **"MRSA"** infection can be quite serious, it may be easily treated effectively with the proper antibiotics. **Can you get MRSA more than once?** The answer to that question is yes! Having had one **"MRSA"** infection doesn't make you immune to future infections, and when treating **"MRSA"** skin infections, doctors often drain the lesion and prescribe antibiotics.

Severe cases can necessitate surgery to remove the infection, the abscess, and the diseased and damaged tissue within the wound. Below is a photograph or a (selfie) of my nose taken by me just three days prior to my hospitalization and surgery. Just look at how fat my face was; it's horrible! ☹

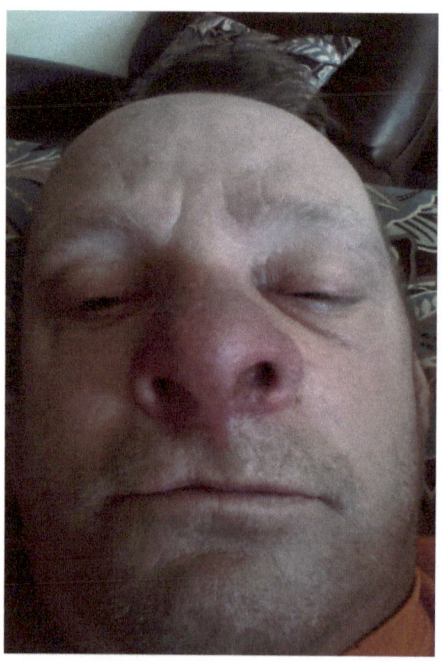

My condition actually worsened after that "selfie" was taken, but my face was so swollen by then, that I couldn't focus well enough to take another photograph. And by that time I had developed what I call **"G.W.P.F.W.B. S.N.A.S."**.

And that my friend simply stands for... Giant, wrinkled, pumpkin face, with big, strawberry nose attached, syndrome. And when people would ask me "what did it feel like"? I would tell them that it actually felt just like someone had sucker-punched me right in the nose with an "SUV" and I think that it was quite possibly a Cadillac Escalade, but it tasted like a Lincoln Navigator. Because that is exactly what it felt and tasted like to me. It was like having a mouthful of metal and medicine for some reason.

I do not know which one it was because I didn't see it sneaking up on me. **"MRSA"** is like a very devious advisory that fights dirty so don't ever turn your back. Below is another photograph takin by me the night before my surgery, and you can clearly see the misery in my poor-fat-face as I lay there in the hospital bed at St. Anthony's Hospital.

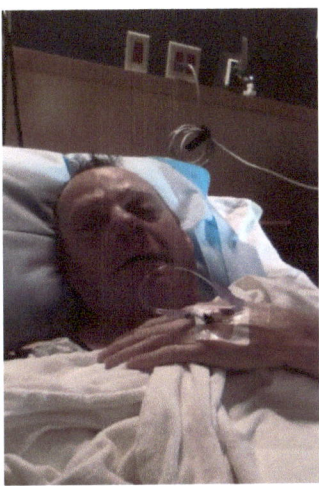

Who can get "MRSA"? MRSA in the community is actually wide spread and therefore anyone and everyone is at risk. People with the greatest opportunity for exposure are those in locations where there is shared space, such as dormitories, military barracks, locker rooms, correctional facilities, elderly care and child day care centers.

In the healthcare setting, those who are more than likely to get infected or colonized with **"MRSA"** are those who are very ill and have an extremely low immune system.

Are in the hospital for a long period of time, have a serious disease that harms the body's ability to fight infection; or have simply taken many antibiotics in the past or present. **What is "MRSA" Colonization?** Some people can pick up and carry the **"MRSA"** on their skin or in their nose for weeks or even months. These people do not get sick or even show any symptoms for that matter; but they do have the **"MRSA"** virus. The bacteria are in fact present but it does not cause any infection.

And as I mentioned earlier, the **"MRSA"** can be colonized in the nose and other body areas such as the armpit and or the crotch area.

Wow! I would truly hate to get a flesh eating bacteria down there my friend. I have already lost about two pounds down there when I was circumcised. I would surly hate to have that hungry, flesh-eating monster, take away any more skin from me down there.

Especially since I'm hung like a new-born baby as it is; sixteen-inches, seven and a half-pounds.

☺**What are the signs of a "MRSA" infection?** An infection can start when **"MRSA"** gets into a cut, scrape, or other break in the skin.

Pimples, rashes, or pus-filled boils, as we mentioned earlier, especially when warm, painful, red, or swollen, can mean a staph or **"MRSA"** skin infection.

Occasionally; staph can cause more serious problems. These problems include high fever, swelling, heat and pain around the wound, headache and fatigue. I actually felt dizzy on several occasions prior to my surgery. **How will I know if I have a MRSA infection or Colonization?**

Lab test are the only way to tell if you have a **"MRSA"** infection, and these tests will help your doctor decide which antibiotic should be used for treatment, if antibiotic treatment is actually necessary. If you have a **"MRSA"** infection your doctor will usually take a sample on a swab (like a Q-tip) from the infected area.

It will then be sent to the lab to see if the infection is caused by **"MRSA"**, and if checking for colonization your doctor will take a sample on a swab inside of the nose, armpit, or groin area. Below is a photo of a typical MRSA infection on the skin.

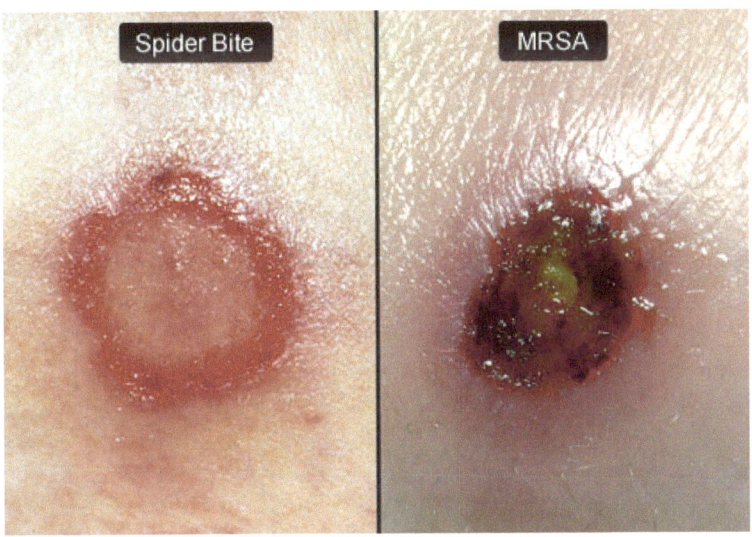

I once had a MRSA infection on my skin which I believed at the time was from a spider bite. Because when I first discovered it I had spent two days (prior to the outbreak) cutting down some old trees that had been dead for many years, and they were infested with spiders, termites and many other insects that I could not identify.

If you think that you have a **"MRSA"** infection, keep the infected area covered with clean, dry bandages. Clean your hands constantly with antibacterial soap, warm or hot water and a scrub brush and spend the extra effort to scrubbing under all of your finger nails extremely well. After changing any bandages, be sure to place all of the old dressings into a plastic bag, tie it closed, and throw it away promptly. If you have been hospitalized from a **"MRSA"** infection and you have returned home, it is ok; just make sure that everyone practices systematic hand washing.

No special cleaning is required as laundry and dishes can be done as usual. And if you ever by chance have **"MRSA"** be sure to tell anyone who may be caring for you that you have it and be sure that you complete all of your prescribed antibiotics to your doctor's directions for proper use. Other types of **"MRSA"** skin infections may look similar to this photograph shown below.

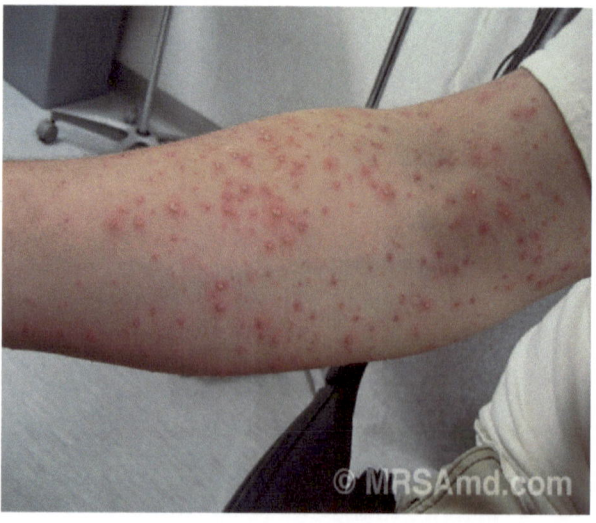

The experts at **"Web-MD"** state that one third of the American population carry staph bacteria in their noses and can cause serious problems like infected wounds or even pneumonia. Staph can usually be treated with normal antibiotics. But over the decades, some strains of staph like **"MRSA"** have become stronger and resistant to antibiotics that once destroyed it.

Health experts now state that **"MRSA"** is now resistant to the following antibiotics; methicillin, amoxicillin, penicillin, oxacillin, and many other common antibiotics.

While some antibiotics still work, **"MRSA"** is constantly adapting to them and growing stronger each time and researchers who are developing new antibiotics are having a very tough time keeping up with it. Below is another microscopic photograph of **"MRSA"** which I acquired from the internet compliments of **"Web-MD"**.

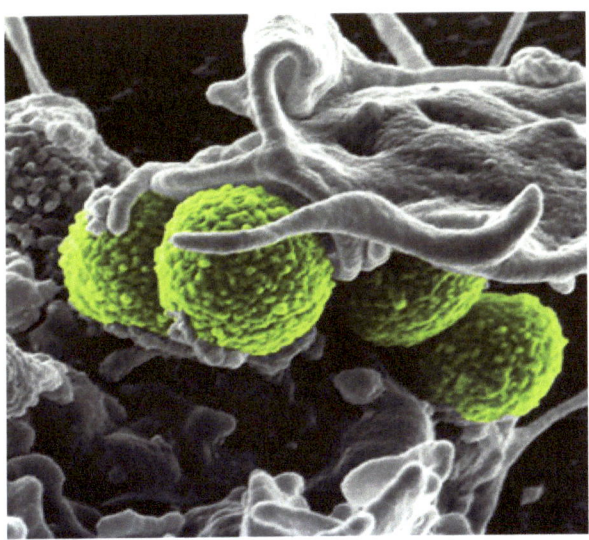

Remember what I said to you earlier, two out of every one-hundred people in the united states carry **"MRSA"** and we should all be practicing simple but yet important hand-washing precautions, and always wash your hands thoroughly for a minimum of twenty seconds.

When Precautions Fail? When all of your **"MRSA"** precautions have failed, get yourself to a doctor or hospital emergency room immediately! Don't waste any time what so ever, as I did foolishly thinking that I was simply suffering from a head cold that turned into a sinus infection. Here is some advice as written by Jude White. He is the co-founder of MRSA.com, an informational website to help others understand more about this "nasty-ass" infection.

When **"MRSA"** precautions have failed, treatment will more than likely be necessary. In addition to treatment, the infected area must be covered at all times and all steps taken to avoid the spread of the bacteria to others who may come in contact with the infected person.

Here is a little information that I discovered from the University of Berkeley in California. Fascination with tiny microbes bearing long and difficult-to-pronounce names is often reserved for biology classrooms, unless of course the bug in question **(MRSA)** threatens human health. MRSA (AKA) the super-bug, now contributes to more deaths right here in the U.S. than does "HIV", and as its threat level has risen, so has the attention lavished on it by the media.

But yet I have never heard of this horrible bacterium from hell until this year. Other findings in this article state that humans and their pets can share this infection with one another, and that makes this "methicillin resistant monster" one bad-mother. Luckily, from what I have been told we still have two "kick-ass" soldiers who can knock the living hell out of **"MRSA"**, and that is **"Vancomycin"** which is often considered our last line of antibacterial defense.

And then there is the other one is **"Clindamycin"** which is the one that they gave to me at the hospital. They first administered it to me through the "IV" in my arm, and after I was discharged I continued taking it orally for an additional ten days afterward.

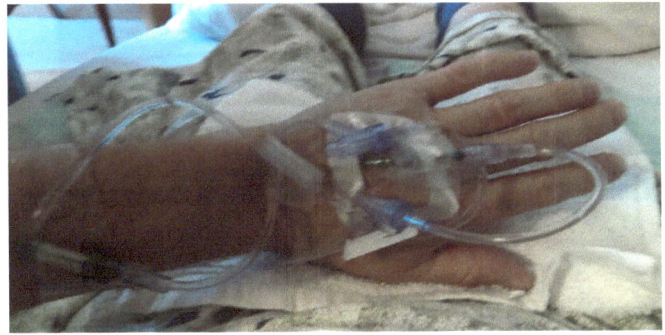

They now call me; the "MRSA-NATOR". After my surgery was completed and I had completely awakened from the anesthesia, I immediately felt an improvement. I still wasn't able to breathe through my nose just yet, but I felt so much better than I did when they rolled me into the operating room that morning at six-am. And you can literally see the difference in my face in the photograph of me shown on the next page.

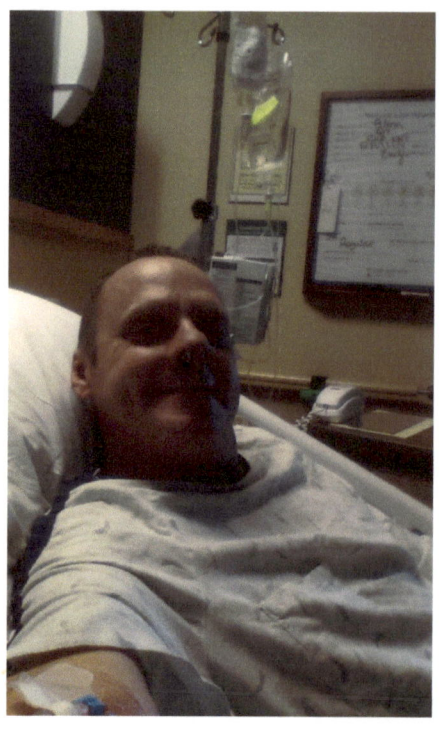

And, if you look closely; you can see the "stents" which were sutured into my nostrils during the surgery. They were placed there to keep my nostrils from not only collapsing but to allow me to finally breathe once again. So, at this point I would like to give special thanks to my surgeon; **"Dr. Keith Ladner"** and his extremely talented crew in the "O.R." that day. And I would also like to personally thank each and every one who cared for me during my stage of recovery at **"St. Anthony Hospital"**.

Thank you all so much! As each and every one of you were all professional and absolutely marvelous! Furthermore, you are all going to get a free copy of this book which I will personally hand-deliver when I get it published; I promise! Now, check out this horrible photograph of my nostrils all scabbed up. This is one of my favorite selfie-pics which I took during my stay at **St. Anthony's Hospital**. This is a total ten on the **"YUK"** factor here!

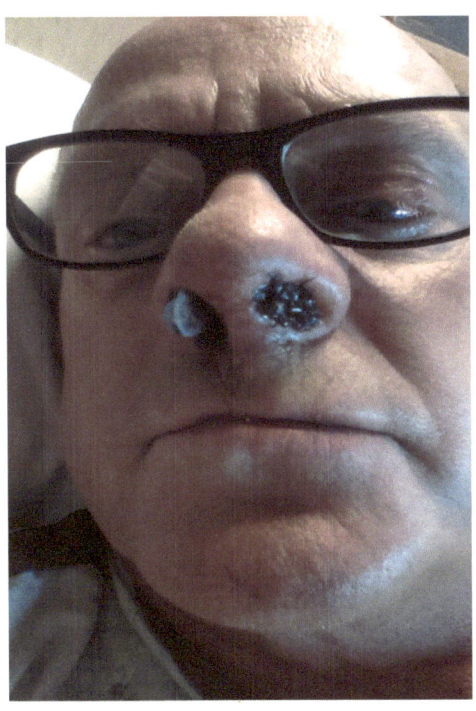

Now that my friend, is some serious **"scabbage"** as I like to call it. I know that there is no such word, but I sure like the way it sounds, and if **"scabbage"** ever existed; it would look just like this. After my second day of recovery I was finally able to take a shower and let the cascading water from the shower-head clean out the scabs from my poor little nostrils.

Then the healing process began and I was then able to do a little more cleaning each day after that. My personal cleaning of my nostrils was pretty gross, and with clean hands and some clean tissue, I was able to twist the tissue into a long and skinny probe which I could slide up inside of each nostril alongside of the stents that were placed inside of my nose. I would then twirl the long, skinny, Q-tip like probe in one direction.

And as I did so I could feel it catching the mucus inside of each nostril, and while doing so I slowly and gently pulled my homemade "snot remover" out of the nostril, and while doing this technique, the discharge literally wrapped itself around the skinny tissue wand looking similar to a "Barber Pole" or a candy-cane. Then, it was easily removed while relieving some of the pressure from all of that sinus build up inside. Oh, God it was so gross! Like a ten on the "gross-o-meter". I had big and thick, long, slimy, yellowish and bloody snots that were as long as a night crawler. If I hadn't been feeling so sickly, I would have saved them in a jar and taken it to the lake where I do my bass-fishing. "Yow". Now that's just nasty, but I like it; so I'm putting it in the book!

☺ In an article, written by "Dr. Robert L. Pincus MD". He states that **"MRSA"** is not especially contagious. The problem is that most doctors treat infections with antibiotics that will not work against **"MRSA"** until the patient gets sicker on the medications. Today, we will often treat this initially with medications that will in fact work against most **"MRSA"**. **"MRSA"** is becoming more and more common in the American community today, even in normal and healthy individuals. If you think that you or anyone in your family may have **"MRSA"** I suggest that you immediately see your health-care provider for a culture.

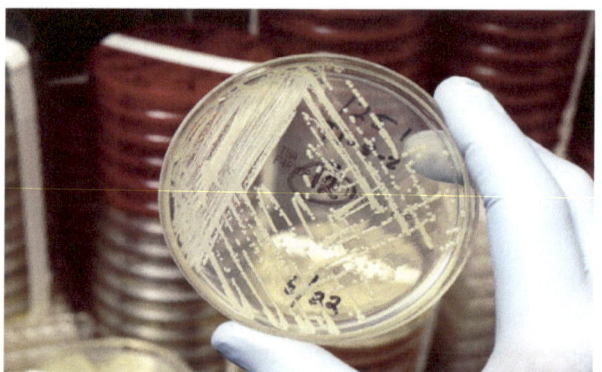

In the year of 2007 a published **"JAMA"** study showed that 95,000 invasive **"MRSA"** infections occurred in the United States in the year of 2005. And that these infections were associated with death in approximately 19,000 cases. That right there is enough to scare the hell out of me, excuse me while I go wash my hands for about thirty-seven minutes or so. There is some good news here however. And that is that the **"CDC"** reported in the year of 2012 that **"MRSA"** cases in hospitals has dramatically declined, and that here in the U.S.A. We are getting a very good grip on the menacing monster and kicking some **"MRSA"** ass!

So, that is some good news to help you from freaking out. Anyway, that is about the extent of what I can tell you about my experience with **"MRSA"** today. So, I will bid you good bye, good luck, and shall we say (hopefully) good reddens to the **"MRSA"** monster that made me feel like I was going to die. And once again I wish to thank everyone at **St. Anthony's Hospital** for all of their professional, compassionate effort, in caring for me during my speedy recovery!

St. Anthony Hospital.
Lakewood, Colorado. November 2016.
This is the best hospital I have ever been in!
And I have tried out many!

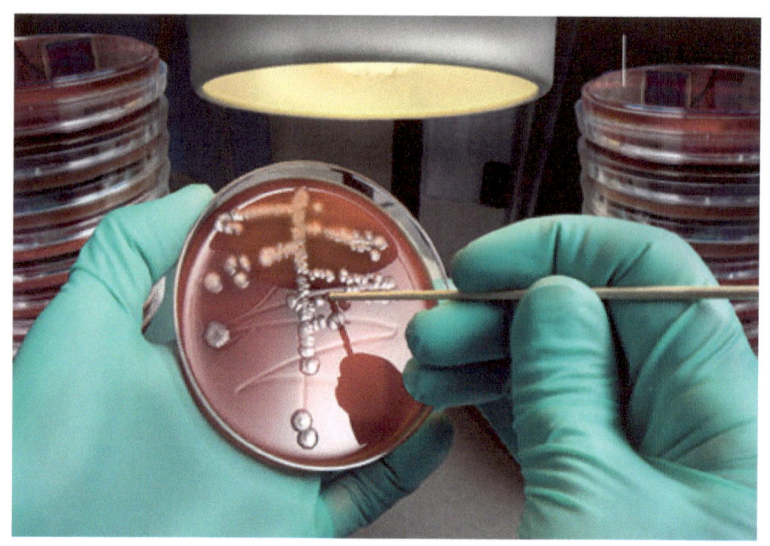

Lab testing for "MRSA".

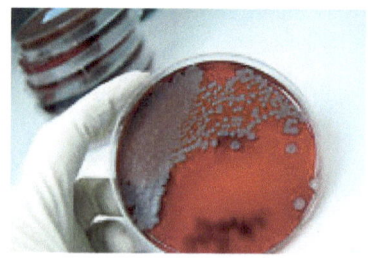

Photographs; compliments of "WEB-MD"

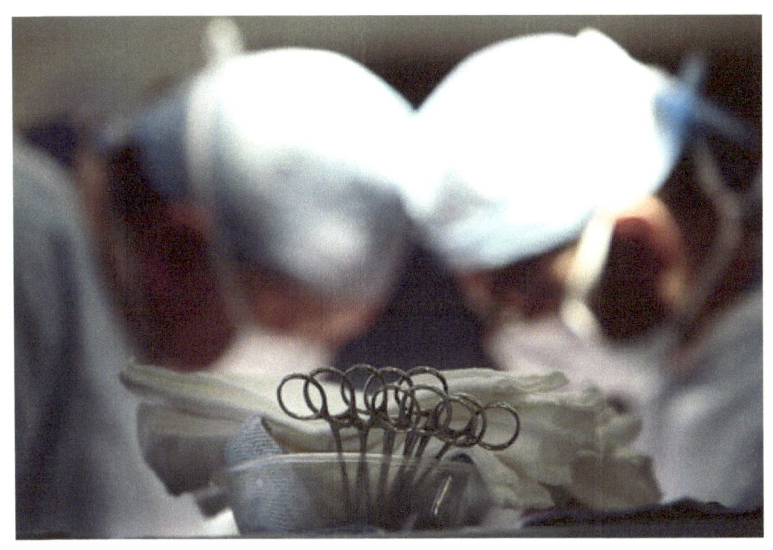

Surgeons hard at work removing "MRSA".

Jimmy "Francis" Durante
Famous for his big nose.
Died 01/21/1980

THE END

Glenn Michael Demarest
Copyright
2016

www.ingramcontent.com/pod-product-compliance
Lightning Source LLC
Chambersburg PA
CBHW041115180526
45172CB00001B/256